Generis

PUBLISHING

GROWING IN HOMES

Two Nutritious Plants, Amaranth and Nopal

EDMUNDO ARIAS TORRES

Title: GROWING IN HOMES

Two Nutritious Plants, Amaranth and Nopal

ISBN: 979-8-88676-408-6

Author: EDMUNDO ARIAS TORRES

Cover image: www.pixabay.com

Publisher: Generis Publishing
Online orders: www.generis-publishing.com
Contact email: info@generis-publishing.com

AKNOWLEDGEMENTS

To Dr. Imilla I. Arias Olguín for her contributions.

To Bolivar M. Arias Olguín for his opinions.

INDEX

ABSTRACT

Amaranth and Nopal plants have been part of the Mexican diet of the inhabitants of semi-arid zones mainly, since ancient times. These two plants contain protein and grow with little water and intense solar radiation. They are usually found in open spaces in the countryside.

These plants have special dietary characteristics.
Amaranth seeds contain between 13 and 18% protein and vitamins A, B, B1, B2, B3 and C, which gives it a high nutritional value. The Amaranth is indicated for celiacs since it does not contain gluten, neither has cholesterol.

Nopal have functional compounds for the body because has an inportant amount of fiber, minerals as calcium and potassium, also some vitamins such as vitamin C.
Due to their nutritional properties it is recommended for children, older adults and specifically for diabetics, neither has cholesterol.
The Nopal has a fruit called prickly pear that like the Amaranth, can also be eaten as a sweet.
Nopal is also used in cosmetics and shampoo.

The two plants can be grown in small spaces, so it is proposed to plant them in houses with small service yards. Its plantation can be in pots and drawers that require few wooden boards to make them.

These products in addition to being protein elements, help in the family economy by reducing expenses on protein of animal origin.

The residues of both plants can be used for small-scale composting.

INTRODUCTION

The socioeconomic situation worldwide continues to be difficult with the pandemic and different political and economic events that have repercussions at a social level. This has been reflected in the increases in food costs, mainly in beef and chicken, which has influenced the fact that people's diets have decreased in quantity and quality.

However, there are alternative agricultural products that contain protein and can be grown in small spaces and included in the diet of nutritious food for people and even small animal species such as chickens and rabbits.

There are two unusual examples of agricultural products that require less water and less demanding soils for their cultivation compared to others, these are: Amaranth and Nopal.

Both plants are related to adverse agroclimatic conditions such as receiving a lot of solar radiation during the day and a downward temperature variation at night, with little water availability.

Amaranth has the potential to become a basic crop of the same agricultural and economic importance as other species such as corn, wheat, sorghum, barley, rice, among others.
Its cultivation has the potential for agronomic and industrial development similar to soybeans.

The Nopal is one of the most important plant resources for its cultivation in semi-arid zones since it resists climatic conditions of intense solar radiation, little water and low temperatures.

It is proposed in this Book to help improve the diet of residents living in places with little free construction space in their houses and take advantage of it to cultivate these two minor species in a sustainable way and use them to improve their family diet, besides can reduce their expenses in buy beef, pork, chicken and rabbit, thus helping their economy.

Other advantages of both plants is that they can be used not only as food for humans but also for cattle and smaller species such as chicken and rabbits. They can also be used to make cosmetic products, the waste can also taken advantage to make compost.

AMARANTH

SOURCE

There are species native to Mexico that are of great importance as food and a symbol of the Mexican indigenous culture, such as Amaranth.

It is an annual herbaceous plant that belongs to the genus Amaranthus, predominantly tropical, including about 70 species native to the tropics and temperate regions around the world; of which 40 are from the American continent and the rest belong to Australia, Africa, Asia and Europe. Within the genus are the species A. cruentus, A. hypochondriacus and A. caudatus, which are the most important for the production of Amaranth grain. The first two are widely distributed in Mexico, forming the most important center of diversity. Sauer (1950, 1967).

According to Grubben (1975) and Grubben and Sloten (1981) Amaranthus cruentus L., is a species for grain production, is native to Central America, probably Guatemala and southeastern Mexico, where it is cultivated and widely distributed.

Amaranth is known for its high biological quality of proteins, vitamin E.
It has wide range of uses in human nutrition and agronomic potential such as resistance to drought and broad adaptation to various environments. All these characteristics allow to postulate that Amaranth is a potential crop to reduce food problems and malnutrition in consumers.

CLASSIFICATION

The Amaranthaceae family is comprised of more than 60 genera and 800 species of annual or perennial herbaceous plants. Only two species of the genus Amaranthus are cultivated for the production of edible seeds (Paredes, Barba 2006).

The best known species are:
Amaranthus cruentus, native to Mexico and Central America (figure No. 1)
Amaranthus hipochondriacus, from the central part of Mexico. (Figure No. 2)

Both species are grown for grain. (Assad 2017)

Fig. No. 1   *A. cruentus*

Fig. No. 2   *A. hipochondriacus*

Table No.1   Taxonomic classification of Amaranth

kingdom	Vegetal
Division	Magnoliophyta
Class	Magnoliopsida
Subclass	Caryophyllidae
Order	Caryophylliales
Family	Amaranthaceae
Genus	Amaranthus
Section	Amaranthus
Species	Amaranthus cruentus
	Amaranthus hypochondriacus

Tuston 2007

Table No. 2 soil characteristics	(%)
Sandy	88,6
Silt	5,4
Clayey	6
pH	Little acid

Tuston 2007

Amaranth is a plant whose central stem can measure up to 2.4 meters (8 feet) in height, however there are some varieties of this same plant with less altitude. It has very showy flowers that sprout from its central stem, with cylindrical branches and a short main root (Fig.3).

This plant has a great capacity to adapt and survive the different environments in which it is grown.

Fig. No. 3 Amaranth in flower

PROPERTIES

Amaranth is a crop alternative for locations where cereals and vegetables cannot be grown due to weather conditions and soils in semi-arid regions (Omami et al., 2006).

Fig. No 4   Amaranth cultivation in the field.

In the 1970s very high levels of protein were found in Amaranth, which according to the Food and Agriculture Organization of the United Nations (FAO), is the ideal protein for human beings; also lysine, one of the 9 essential amino acids (National Research Council, 1984).   One of its strategies is the promotion and incorporation of Amaranth in the diet of the population (Morales et al, 2009).

Amaranth contains between 16 and 17% protein, this percentage being higher than that of wheat (12-14%), rice (7-10%) and corn (9-10%).
Various authors indicate that Amaranth can help combat anemia due to the iron and vitamin C found in the leaves and seeds.   It also reverses malnutrition related to the consumption of low-protein foods, since it contains an almost perfect balance of amino acids and abundant lysine; it is excellent compared to other foods of animal or vegetable origin such as meat, milk, eggs, corn, beans and others.

Amaranth leaves can also be consumed due that contain essential vitamins and minerals such as calcium, phosphorus and folic acid. They have more iron than spinach, so it is recommended for people who suffer from a certain degree of anemia, especially among women and children.

For all of the above, Amaranth has been considered by the National Academy of Sciences of the United States (NAS) and by the FAO as the most promising plant for agricultural and nutritional development; especially for semi-arid and arid regions, where agriculture is seasonal. It can also be an option as an alternative and integral crop, where it is produced, consumed and marketed, resulting in a decrease in malnutrition, generation of economic income and family well-being at the community level.

Table No. 3   Composition (%) Raw amaranth

Water	12
Carbohydrates	7
Protein	7
Fat	7

Tuston 2007

As can be seen in table No. 3, Amaranth seeds are a source of protein, they also contain several dietary minerals that are retained even when cooked. They also have: calcium, magnesium, phosphorus and potassium.
In addition, cooked amaranth leaves are a rich source of vitamins A and C, calcium, and manganese, with moderate levels of phosphate, iron, magnesium, and potassium.
An important feature is that they do not contain gluten.

Point out that it is an easy-to-manage crop, since it adapts to diverse environmental conditions.   Farmers recognize that they maintain the crop because the seed is nutritious and they are interested in their families consuming it.
Some of them who cultivate it confirm that it helps the family union by participating several or all the members in the processes of production and transformation of Amaranth products

Table No. 4   Composition of essential amino acids of seeds of two species of Amaranth in (g/100g of protein)

AMINO ACID	Amaranthus hypochondriacus	Amaranthus cruentus	FAO/WHU/UNUA Adult's	children
Isoleucine	2.8 – 3.8	3.4 – 3.7	1.3	4.6
Leucine	5.0 – 5.8	4.8 – 5.9	1.9	9.3
Lysine	3.2 – 6.0	4.8 – 5.9	1.6	6.6
Met + Cisb	2.6 – 5.5	3.8 – 5.4	1.7	4.2
Fenila + Tiroc	6.9 – 8.5	5.6 – 8.5	1.9	7.2
Threonine	2.6 – 4.3	3.2 – 4.2	0.9	4.3
Tryptophan	1.1 – 4.3	nd	0.5	1.7
Valine	3.2 – 4.2	2.4 – 4.0	1.3	7.2

Silva 2007.

Table No. 5   Approximate composition of Amaranth and spinach leaves (100mg)

Component	Amaranth	Spinach
Protein	3.5	3.2
Calcium	0.262	0.093
Phosphorus	0.067	0.0519
Iron	0.0039	0.0031
Vitamin A	6100	8100
Ascorbic acid	0.080	0.051
Humidity	86.9	90.7

Matías L et al 2018

Table No.6   Approximate composition of Amaranth and the main cereals.

Component	Amaranth	Corn	Rice	Wheat
Protein	17.9	10.3	8.5	14.0
Fiber	2.2	2.3	0.9	2.6
Carbohydrates	57.0	67.7	75.4	66.9
Fat	7.7	4.5	2.1	2.1
Humidity	11.1	13.8	11.7	12.5
Ashes	4.1	1.4	1.4	1.9

Mathias L et al 2018.

CROP.

PRECOCITY.   The duration of the life cycle of the Amaranth plant can vary from one region to another, from one race to another, in such a way that the classification by stages: early, intermediate and late, cannot be applied without the reference of environmental conditions.
According to the evaluation of genetic variability, Amaranths grown for grain vary from 70 to 240 days to mature, depending on the species, race and genotype.   The difference is both genetic and environmental, especially influencing factors such as photoperiod and temperature.

TEMPERATURE.   Amaranth develops best when daily temperatures reach at least 21 °C (69.8 °F).   Many plants have shown optimum germination when the temperature is 16 to 35°C (60.8 to 95°F). (National Research Council, 1985). For its development, Amaranth requires at least 120 days with temperatures above 12 °C (53.6 °F); with 8 °C (46.4 °F), it stops growth below 4 °C (39.2 °F) and suffers damage.

ALTITUDE.   It grows satisfactorily from sea level to 2,500-3,200 meters (8,202-10,498 feet), so it is not a severe limitation of Amaranth for its adaptation.   A Caudatus is the only species cultivated at altitudes greater than 2,500 meters above sea level (masl),(8,202 ft). (National Research Council, 1985). (Espitia et al. 2010) report that A. Cruentus is mainly concentrated in a range from 1,000 to 2,400 (masl), (3,280 to 7,874 feet).

I USUALLY.   Amaranth requires well-drained soils and develops best in neutral or basic soils, pH above 6.

GROWING AT HOME IN SERVICE YARDS

Planting can be done in a pot or drawer placed in the yard of a house.   A 1cm hole deep is made in moist soil and eight seeds are placed.   The humidity of the earth is enough not to water it for five days.
Sowing can also be done in drawers with groups separated by 20 cm (7.87 inches), depositing 10 to 20 seeds for each group.
Moisture is needed only at the time of sowing, until the sprouts appear.

The total amount of water required by Amaranth throughout its life cycle is only 60% compared to wheat or barley.

Its vegetative cycle has an average of 180 days, from when it germinates until the seed reaches maturity. It is important to have the plant in a sunny place.

Three months later the plant reaches maturity and a height of approximately two meters.

Harvest flowering occurs four months after the seed was planted. When the seeds are dry on the stem they can be harvested the same day, if not, they are left to dry for two to three days. Since they are dry, can be separated from the stem through a sieve.

Fig. No. 5 Cultivation in pot.

Fig. No. 6 Cultivation in drawer.

PRECIPITATION.

For Amaranth seeds to germinate they need good soil moisture; once seedlings are established, they can grow well even with limited water. In fact, Amaranth grows best under conditions of low moisture availability and high temperatures (National Research Council, 1985).

It develops adequately with annual rainfall of 469 to 1,347 millimeters (mm), distributed mainly from June to October (Reyna, 1986), in Mexico. However, it has been reported in regions with annual rainfall of up to 200 mm (FAO, 1994). The most acceptable annual rainfall for the crop is the one that ranges between 400 and 1000 mm; however, this species can be established from 300 to 2000 mm.

There is excellent potential in low rainfall areas, which have traditionally been planted with sorghum and millet (Kauffman and Hass 1984, Kauffman et al., 1984 and Weber et al., 1985).

APPLICATIONS.

The entire Amaranth plant can be used in its entirety. The leaves can be consumed as a vegetable and the stem can serve as stubble or fodder for animals.

Amaranth has been used as a pole for the development of small communities where there is a shortage of water, because it is a short-cycle crop so that the land is worked and not left abandoned. Some farmers indicate that they use the seed as raw material for their agribusiness; Others comment that they have insurance in the market. (Cortes et al., 2010).

TECHNOLOGICAL APPLICATIONS.

Amaranth has two types of starch: binder and non-binder, the first is the most suitable for the baking industry and is the one found in some cereals such as rice, corn, barley, sorghum and millets; Amaranth has the first characteristic to be used in this industry (Okuno and Sakaguchi, 1984); however, it can be used in the production of baked goods that do not need expansion because it lacks functional gluten and can be used in mixtures with other cereal flours (National Research Council, 1984). In this regard (Lorenz 1981) points out that Amaranth can be used in the preparation of breads in substitution of 10% of wheat flour, which improves the nutritional quality and flavor is described as very nutty and was preferred over bread made with 100% wheat flour.

Amaranth can be used in the form of refined white flour or wholemeal flour in various bakery products such as: cookies, pancakes, cakes, and bread. The breads with a 10% substitution are of very good quality and improved in flavor compared to the breads with 100% wheat; in the case of biscuits, these support very high percentages of substitution and in all cases improve the characteristics of wheat biscuits; regarding the preparation of pancakes, good results were obtained when using both refined and wholemeal Amaranth flour, the texture is not affected, which must be soft and fluffy, and in the case of pastries, when the substitution was 10 and 20%, obtained a clear improvement with both refined and wholemeal flours.

Bressani and Elías (1984) indicate that there are many ways of consuming Amaranth, but most of them are in mixtures or using it as a complement; They mention that the best option is to use it in products that are made 100% with Amaranth in order to make the most of its nutritional properties.

In Mexico, a basic element is the corn tortilla. (Castilla et al. 1982) found that up to 10% of Amaranth flour can be added to nixtamalized corn flour without altering the characteristics of the product. When the Amaranth flour is from roasted seed, an improvement was noted in terms of flavor, consistency and acceptance of the product.

Amaranth can also be used in a wide variety of products, for example: soups, pancakes, breakfast cereal, buns, crepes, toast, tortillas, fried foods, cookies, empanadas, pasta, snacks, beverages and confectionery (National Research Council, 1984).

AMARANTH IN THE INDUSTRY.

The use of Amaranth in the pharmaceutical and cosmetic industry has increased considerably over time, because the oil is rich in squalene, it is used in these two branches of the industry.

On the other hand, Amaranth oil is not particularly unique, it is very similar in composition to cotton and corn. However, recent studies have found a relatively high content of squalene (approx. 7 to 8% of the seed oil). This substance is an important ingredient in the pharmaceutical industry as a precursor to steroids and in the cosmetic industry, and lubricant for machines.

Amaranth can be industrialized to obtain modified starches, drugs, edible oil, protein concentrates and cosmetics. Therefore each of these products can be inserted in many of the important markets.

Fig. No.7 Amaranth industrialized products

RECIPES WITH AMARANTH.

BREADED FISH WITH AMARANTH

Ingredients
- 4 skinless fish fillets
- 1 teaspoon of pepper
- 1 ½ teaspoon salt
- ¼ teaspoon garlic powder
- ½ teaspoon onion powder
- ½ teaspoon paprika powder
- ¼ cup of rice flour
- 1 egg
- 1 ½ cup amaranth
- 1 cup vegetable oil

Preparation.

1. Mix salt, pepper, garlic, onion and paprika powder.
2. Season the fish files on both sides.
3. Cover each one with rice flour,
4. Pour the egg into a container and pass the fillet on both sides in the consistency and lastly in the amaranth.
5. It can be prepared in two ways: 1. By heating vegetable oil and frying the fillets.
2. If you prefer, place a rack on a baking sheet, arrange the fish fillets on top, spray them with a little spray oil and bake them at 180°C for 10 minutes.
6. Remove from the oven and serve with steamed vegetables.

AMARANTH PANCAKE

Ingredients

- 2/3 cup of flour
- 1 cup of milk
- ½ cup vegetable oil
- ¾ cup of refined sugar
- 5 eggs
- 2 teaspoons of vanilla essence
- 4 cups of amaranth
- 1 tablespoon of baking powder
- ½ cup of blueberries
- ¼ cup of finely chopped walnut
- ¼ teaspoon of salt

Preparation

1. Place the sugar, oil, milk, flour, baking powder, vanilla essence, salt and eggs in a blender; blend until everything is well incorporated.
2. Mix the amaranth with the blueberries and the blended preparation; beat until everything is well incorporated.
3. Grease and flour a baking pan, empty the mixture until filling 2/3 parts of the mold.
4. Sprinkle the walnuts on top and bake at 180°C for 40 minutes or until, when
5. insert a toothpick in the center, it comes out clean.

AMARANTH ATOLE

Ingredients

- 1 liter of milk
- 250 gr. (8.8 oz) amaranth
- 2 cinnamon sticks
- 1/3 cup of sugar

Preparation

1. Heat the milk with the sugar and cinnamon over medium heat, let it boil.
2. Add the amaranth when the milk begins to bubble. Cook for 10 minutes.
3. Let cool and blend the mixture so that it is free of lumps.
4. Heat over low heat until it takes a thick consistency.

CANDY WITH AMARANTH. (ALEGRÍA)

Ingredients.

6.6 lb. of amaranth
5.5 lb. of sugar
5 cups of water

Preparation

1. Wet the Amaranth and drain. Let dry it in the sun or in the oven at a very low temperature.
2. Toast the Amaranth in pan over regular heat.
3. Clean to loosen all the husk.
4. Add the sugar with the water over medium heat. Stir until dissolved and let boil.
5. Remove from the heat and add the Amaranth, mixing to form a paste.
6. Use galvanized metal molds, preferably 3 cm wide.
7. Arrange them on a clean and damp table and serve the mixture in the molds.
8. Take them out and let them cool completely.
9. Serve the mixture in the rings and press with circles.

CONCLUSIONS.

Amaranth is an agricultural product suitable for semi-arid areas where crops are scarce due to lack of water.

Its nutritional properties have been highlighted by International Organizations such as the FAO due to its protein content. It also contains calcium, magnesium, phosphorus and potassium.

It is a food that can be used in its entirety since the leaves are a rich source of vitamins A and C, containing higher levels of iron than spinach.

An important characteristic is that neither the seeds nor the leaves contain gluten, which makes it a beneficial food for celiacs.

It is very important to highlight that in addition to improving the diet of those who consume it, they also improve their economy by not having to spend on beef, pork, or chicken to eat protein.   They can also prepare stews or desserts to eat or sell.

In addition, Amaranth has applications in the pharmaceutical and cosmetics industry.

N O P A L.

S O U R C E.

Opuntia spp., popularly known in Mexico as Nopal, is a plant that belongs to the cacti, which due to its characteristics is ideal for growing in semi-arid and arid areas.

There is evidence in archeology that affirms that it was the indigenous populations settled in the semi-arid zones of Mesoamerica that began its cultivation in a formal way.

The Nopal is one of the most important food resources of the Mexican flora. Currently due to its various nutritional, chemical, industrial, ecological, medicinal and symbolic properties, among others, the Nopal is one of the most important plant resources for the inhabitants of the arid and semi-arid zones of Mexico.

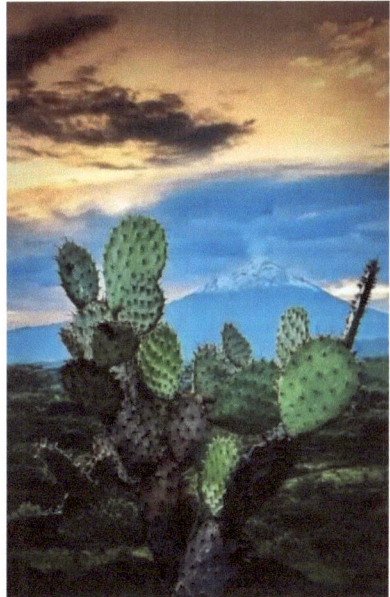

The Nopal is linked to the history of Mexico in a particular way, as can be seen in the shield of the Mexican flag where there is the figure of an eagle perched on a Nopal.

It is native to tropical and subtropical America; Over time it has managed to spread in Africa, Asia, Europe and Oceania, where it grows wild or is cultivated.

F E A T U R E S.

The Nopal is a fleshy and thick plant with diverse articulated branches, it has leaves turned into thorns which is a common characteristic in cacti.-
This plant has a very thick cuticle, almost all with spines; stores large amounts of milky or rubbery juice (mucilage) that allows it to resist drought; (Acevedo et al., 1983)

Fig. No. 1 Nopal

It presents the metabolism of crusalaceous acid (MAC), that is the stomata capture the CO_2 that is used for the synthesis of carbohydrates during the night, allowing less water loss in the sunniest hours of the day. On the other hand, due to its ecophysiology: asynchronous reproduction and structural adaptations (low stomatal density and thick cuticle), the plant is capable of surviving long periods of drought. The Nopal is also used in reforestation programs, due to its ability to grow in poor soils inappropriate for other crops. It grows in environments with extreme temperatures, and the presence of scarce rainfall; it adapts to variations in atmospheric CO_2 levels. It also plays an important role in the ecosystem since it protects wildlife. There are about 300 species of the genus; about 100 of them exist in Mexico and of them, about 40% are located in the desert of the State of Chihuahua, in the North of the Mexican Republic. (Méndez Llorente et al., 2008).
In addition to the characteristics of the Nopal that have been described, it bears a fruit called prickly pear, species Opuntia fiticus-indica. It is an oval berry with an average weight of 0.22 to 0.44 lb. The pulp is juicy and constitutes 60-70%

of the total weight of the fruit; It contains numerous hard and small seeds, which vary between 100 to 400 per fruit.

Fig. No. 2 Nopal with Prickly Pears

CLASSIFICATION.

The genus Opuntia in Mexico has five subgenera, seventeen series and 104 species (Bravo, 1978):

• Subgenus Cylindropuntia includes eight series and 29 species, of which only three are used as forage; O. fulgida, O. cholla and O. imbricata.

• Subgenus Grusonia, has a single species.

• Subgenus Corynopuntia includes eight species.

• Subgenus Opuntia involves 17 series and 63 species of which it is used for fodder:
   O. decumbens, O. microdacys, O. rastrera, O. azurea, O. lindheimeri, O. cantabrigiensis, O. durangensis, O. leucotricha, O. robusta, O. stenopétala, O. rufida, O. violacea, O. phaecantha , O. neochrysacantha and O. pailana.

• Subgenus Stenopuntia with three species, two of which are used for forage: O. stenopétala and O. grandis.

The Nopales that are used as vegetables are three species: O. ficus indica, O. robusta and O. leucotricha.
Five species are used for their fruit: O. hyptiacantha, O. streptacantha, O. megacantha, O. xoconostle, and O. ficusíndica.

The Nopalea genus in Mexico has 10 species (Bravo, 1978), of which probably only one, N. cochenillifera, is used as a Nopal vegetable. In sum, of the 104 species of Opuntia and 10 of Nopalea, five are used for their fruit and four as vegetables (three of Opuntia and one of Nopalea) and 15 species for fodder.
It is considered that this classification is useful for the wild Nopales and for the species used as fodder; however, it is not very useful for the varieties cultivated to produce vegetable Nopal or prickly pears.

Table No.1	Taxonomic classification of Nopal
kingdom	Vegetal
Subkingdom	Embryophita
Division	Angiospermae
Class	Dicot
Subclass	Dialipetalas
Order	Opuntial
Family	Cactaceae
Subfamily	Opuntioideae
Tribe	Opuntiae
Genders	Opuntia and Nopalea

García 2003

P R O P E R T I E S.

Nopal can be considered as a functional food. At present, the general trend in food consumption is to seek those that provide nutrients and that are also beneficial for health and for the prevention of diseases.

Nopales have functional compounds for the body, both the fruits and the cladodes are an important source of fiber, hydrocolloids (mucilages), pigments (betalains and carotenoids), minerals (calcium, potassium) and some vitamins such as C, which has antioxidant properties; These compounds are highly valued for providing good nutrition and health, as well as ingredients for the design of new foods. The contents of these compounds are different in fruits and cladodes, the pulp of the fruit is the richest part in vitamin C while the stems are rich in fiber. The pigments are only found in the fruits and may be present in the skin and pulp.

The Nopal, like other vegetables, contributes a high proportion of water to the diet and is highly valued for its high fiber content, it is also low in lipids and carbohydrates.
Its content is comparable to that of several fruits and vegetables, including: spinach, artichoke, chard, eggplant, broccoli, radish, mango, melon, apricot, grape and others.

Table No. 2   Chemical composition of 100 g of Nopal

Components	Quantity
Carbohydrates	2,86 g
Proteins	1,45 g
Lipids	0,21 g
Cellulosics	3,77 g
Calories	19,95 Units
Vitamin A	0,41 mg
Thiamine	0,03 mg

Riboflavina	0,03 mg
Nicotinic acid	0,32 mg
Ascorbic acid	10,76 mg
Calcium	130,00 mg
Phosphorus	21,00 mg
Iron	1,95 mg

Trujillo 2009

Table No. 3   Nopal Nutritional Composition

Sodium	5mh
Potassium	220mg
Vitamin C:	14mg
Vitamin A:	43mg
Vitamin B09:	6mg

Trujillo 2009

NUTRITIONAL VALUE OF THE TUNA.

In addition to being a sweet and tasty fruit, prickly pear has good nutritional value.

Each one has a variety of minerals and vitamins, among which carotenoids, vitamin C and about 30 calories stand out.

The nutritional composition of prickly pear has been studied extensively. The main components are sugars, fiber, mucilage and pectins, those with less presence are: proteins, amino acids, vitamins and minerals (Kossori, et al. 1998)

Table No. 4   Nutritional composition of the Tuna

Kingdom	Plantae
Division	Magnoliophyta
Class	Magnoliopsida
Order	Caryophillales
Family	Cactaceae
subfamily	Opuntioideae
Gender	Opuntia
Species	*ficus-indica*
binominal name	*O. ficus-indica*

Trujillo 2009

Table No. 5   Prickly Pear Taxonomy

Protein	1g
Fiber	3,6g
Fats	0,4g
Carbohydrates	7.1
Calcium	80mg
Sodium	5mg
Potassium	220mg
Vitamin A	43mg
Vitamin C	14mg
Vitamin B	09.6mg
Calories	40

Trujillo 2009

C R O P.

The planting of the Nopal vegetable is preferably done before the rainy season, since the excessive humidity of the soil and the climatic conditions optimize the development of fungi and bacteria, generating serious damage to the propagule.
The most recommended period to carry out the sowing is in the dry season and with dry weather, which occurs at the end of February until April (in Mexico).
Another period in which the plantation can be carried out is between the months of August and September (in Mexico), although it is not highly recommended, since early frosts can occur and thus damage the plant.
Before sowing begins in the open field, it is necessary to prepare the land where the definitive planting of the Nopal vegetable will be carried out.   To do this, the land is cleaned, the stones are removed, weeding is done and the clods of earth that may be found are crumbled, until the space where it will be planted is leveled to the maximum.

The traditional sowing of the Nopal cladode is carried out as follows:
• Distance between rows from 70 cm to 1.5 m. (2.3 to 5 feet).
• Distance between plants of 30 to 50 cm. (1 to 1.6 feet).
• Depth of the stump is 10 to 20 cm, (0.3 to 0.7 feet), burying a quarter or half of the cladode.

Once the cladodes have been planted, the organic waste should be applied two or three months later, since by this time the cladode already has roots that absorb moisture and nutrients from the soil, which allows the growth and development of them.

The plantation should have a North-South orientation, in which the cladodes will face East-West, since it has been shown that in this way there is greater photosynthetic efficiency and greater root development.

For places where the climate has higher temperatures, the orientation can be East-West or Northwest-Southwest, so that the sun's rays do not fall directly on the surface of the cladode and thus avoid burns.

CULTIVATION AT HOME LEVEL.

To cultivate Nopal, it is necessary germinate seeds from the fruit or propagate a new plant from an existing one.

Grow Nopales from seeds.

Take a small garden pot with a hole in the bottom. Cover the bottom of the pot with a layer of small stones to allow the water to drain better.

Fig. No. 4 Pot preparation

Fill the pot with soil that is half soil and half sand, coarse stone, or loam. Such soils drain better than those with a high clay content and more closely resemble the natural desert soils favored by a cactus.

To plant the seeds, place one or two on top of the soil, gently press them into the soil, and cover them with 1/4 inch (0.6cm) of soil.

Add a little water to moisten the soil.

As the seeds grow, keep the soil moist until they germinate. Re-water the soil until it feels dry when you insert a finger into the soil.

Nopales grown from seed tend to take longer to grow than propagated plants and the resulting cacti could take three or four years to produce flowers and fruit. However, growing plants from seed is important to ensure genetic diversity.

Figure No. 5 Nopal in a pot.

Propagate Nopales.

Another way to grow Nopales is to use a cutting from a cultivated plant.

To propagate Nopales from existing plants, use cuttings from the cladode, these are the flat green, spiny part that makes up the majority of the plant. (Fig. No. 6).

Select a healthy cladode that is medium or long in size, that is between one and three years old to be cut. The ideal is to find a cladode that is free of damage, stains or deformities.

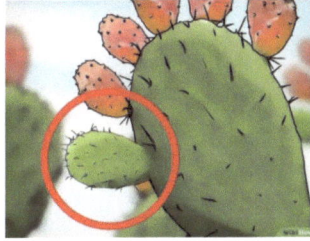

Figure No. 6   Nopal Cutting

Do not cut the stalk below the joint because it can cause an infection and the plant will rot.

To prevent infection and rot, you should let the cutting form a callus where it is cut before planting it.   Place the cladode in a bed of soil or sandy soil for one or two weeks, until the cut has closed.
Leave the stalk in a shaded area while waiting for the callus to form.

Fig. No.7   Scheduling.

For planting the cladode in a pot, fill the bottom with stones to allow water to drain.   Fill the rest of the pot with a mixture of sandy, loamy and clayey soil, with added organic compounds
The ideal soil should be made up of a mixture of half dirt and half sand or stone.

Plant the cladode when the cut has healed. Dig a 2.5 to 5cm (1 to 2 inch) hole in the ground. Don't bury it deeper than suggested, otherwise it could rot.   Stand the cladiode upright with the cut end on the ground or inside the pot.
If the cladode has trouble staying vertical, surround it with a few stones to support it.

Fig. 8 Cultivation of Nopal in soil   37

Do not water the cladiode for a month after planting it, as it will have enough moisture to sprout new roots.

Place the cladiode so that it is sunny. Unlike Nopal seeds, cladiode need a lot of direct sunlight.

To avoid having to constantly move the Nopal, it is recommended to place it so that the wide sides of the pad face East and West. Thus the thinnest sides of the cladiode will be oriented towards the sun when it is at its maximum point of radiation. This will prevent the thinnest part from burning.

After three or four weeks, the cladiode will have lost its moisture and will begin to look dry and wrinkled. This means that it has grown and established new roots and it is time to water it.

APPLICATIONS.

The potential of the Nopal is expanded by being able to be used in different ways, as a vegetable for food; whether in salads, soups, various dishes, tortillas or juices.

When industrialized, they can be canned in brine or pickled.

Nopal waste can be used as an input for homemade composters that can be easily made. Small diameter holes are drilled in a wooden cylinder over the entire surface to aerate its contents. This is placed on a stand that allows you to rotate it every day. It is the way to make homemade compost to fertilize the farmland. The time may vary depending on how late the decomposition process takes due to weather conditions. (Fig.No.9).

Fig. No9   Wooden drum type composter with wire mesh.

It can also be used for medicinal use given the properties of Nopal and its high fiber content, which is why it is used as a natural hypoglycemic to control diabetes mellitus; helps control obesity and lowers cholesterol levels. There are also several reports on the use of Nopal leaves to mitigate pain and inflammation (Sánchez 1982). It is a popular remedy to heal wounds, ulcers and gastritis.

It also has industrial application for obtaining textile dyes; in the cosmetic industry it is used to make shampoos, facial creams, soap, cleansing creams and moisturizing creams, among other products.

Fig. No.10 Moisturizer cream

The potential of the Nopal is expanded by being able to be used as fodder, reducing the pressure on natural grasslands.   It is important because it can be used during times of drought as livestock feed, mainly in arid or semi-arid areas.

Fig. No.11   Jar with prickly pear jam.

The prickly pear in its natural form is consumed as a fruit by removing its covering called the shell, it can also be processed to become jam, jelly, liquor, etc.

The peels can be used in a home composter to which food waste and other fruit peels can be added. (National Commission for the knowledge and use of Biodiversity MEXICO 2009)

RECIPES WITH NOPALES.

NOPAL CREAM

Ingredients
- 8.8 oz (250 gr.) of Nopales cut into pieces
- 1 bunch of spinach
- 1.4 oz. (40 gr.) of butter
- 8.8 oz. (250 gr.) of cream
- 4 cups of water
- 1 teaspoon of chicken broth
- 1 tablespoon chopped coriander
- 1/2 chopped onion

Preparation
1. In a saucepan over low heat, add the water along with the chicken broth, the Nopales, the onion and the butter.   Cook for 20 minutes until the Nopales are very well cooked.   Remove from heat and reserve.
2. Put the spinach in the blender together with the previous preparation, the cream and half of the coriander.   Blend well and return to heat for approximately five minutes.
3. Serve and garnish with a little onion and the remaining coriander

NOPALES STUFFED WITH COTTAGE

Ingredients
- 12 chambray Nopales
- 10.6 oz. (300 gr.) of cottage cheese
- diced onion
- 2 tablespoons of oil
- 2 tablespoons of chopped fragrant herb
- 4 tomatoes
- 1 onion
- 1 clove garlic
- 1 chili in marinade

41

• Salt and pepper to taste

Preparation
1. Put the oil in a frying pan together with the onion and let it fry. Add the cottage cheese and fragrant herb . Season with salt and pepper. Booking.
2. Cook the Nopales in a pan with a little oil.   Salt, pepper and let them roast on both sides.
3. To prepare the sauce, place the tomatoes, garlic and onion on a griddle, let them char (that is, the skin turns slightly black), peel them and take them to the blender along with the chili pepper.   Blend until you get a fine sauce.
4. Fill the Nopales with the cottage cheese and roll up.   You can put a toothpick on them so they don't fall apart.
5. When serving, pour a bed of sauce on the plate and place the rolls on top.
STEAK STRIP WITH ROASTED NOPALES

Ingredients

• 8.8 oz. (250 gr.) lean beef steak
• 2 nopales
•1 clove garlic
• 2 bunches of spring onions
•Salt and pepper
• 8.8 oz. (250 gr.) lean beef steak
• 2 nopales
•1 clove garlic
• 2 bunches of spring onions
•Salt and pepper

Preparation
1. Cut the meat into strips, salt and pepper.   Finely chop the garlic and slice the Nopales into strips.
2. In a frying pan, lightly fry the garlic. When it is transparent, add the Nopales. When it changes color, incorporate the meat.   Continue cooking until well cooked.
3. Clean the onions.   Roast them until they are well browned.
4. Put them together.

CONCLUSIONS.

The Nopal has the great advantage of being able to be cultivated both in pots and in soils of semi-arid climates.

It can be said that the Nopal is a "double" food since both the cladode and the fruit that is the prickly pear can be consumed.

In addition to being nutritional food, it is also medicinal since it is widely used by people with diabetes.

Both the Nopal and the Tuna have been industrialized for food, pharmaceutical and cosmetic products.

Nopal waste and prickly pear shells can be used to make compost in home-type composters.

GENERAL CONCLUSIONS.

Both Amaranth and Nopal are two crops with low water requirements, easy to grow in small spaces or in large areas of land. Both can be grown in semi-arid soils with climatic conditions of intense solar radiation during the day and low temperatures at night.

Likewise, they can be grown in houses with little space, since they can grow in pots.

Both Amaranth and Nopal have nutritional and human health benefits either naturally or as medicines.

These products, in addition to being protein elements, help in the family economy by reducing expenses on protein of animal origin.

The waste from both plants can be used for small-scale composting.

R E F E R E N C E S.

1. Sauer Jd. 1950. The grain amaranthus. a survey of their history and classification. annals of the Missouri botanical garden.

2. Grubben g.j.h. 1975. Culture of the amaranth, a tropical leaf vegetable, with special reference to south dahomey. medelelingen landbouwhogescholl wageningen 75 (6) 223.

3. Grubben g. j. h., Van Sloten. 1981. Genetic resources of amaranths. international board for plant genetic resources. Rome, Italy.

4. Paredes b. López o., F. Guevara Lara, L. A. Bello-Pérez. 2006. Los alimentos mágicos de las culturas indígenas mesoamericanas. la ciencia para todos 212. Secretaría de Educación Pública & Fondo de Cultura Económica. México.

5. Assad, r., Reshi, Za, Jan, s. and Rashid, i. 2017. "Biology of amaranth". the botanical review, 83(4), 382-436.

6. Tuston, s. 2007. Adaptation of five lines of white grain amaranth, amaranthus caudatus and five lines of attack., Universidad Técnica del Norte. Faculty of Agricultural Engineering. San Luis Potosi.

7. Omami, e.n., Hammes, p.s., & Robbertse, p.j. 2006. Differences in salinity tolerance for growth and water-use efficiency in some amaranth (amaranthus spp.) genotypes. New Zealand Journal of crop and horticultural science. volume 34, issue 1.

8. National Research Council 1984, Amaranth: modern prospects for an ancient crop, Washington, D.C., National Academy Press.

9. Morales, G.J.; Vazquez, m.n.; Bressani, C.R. 2009. Amaranth: physical, chemical, toxicological and functional characteristics and nutritional contribution. National Institute of Medical Sciences and Nutrition Salvador Zuribán (incmnsz). Mexico, 269 p., 2009, isbn: 978-607-7797-00-5.

10. Silva Sánchez 2007. Physicochemical Characterization of Nutraceutical amaranth (amaranthus hypochondriacus) cultivated in San Luis Potosí, México.

11. Matias L. et al 2018 Current and potential uses of amaranth (amaranthus spp.)

12. Espitia Rangel E., Mapes Sánchez C. 2010. Conservation and use of genetic resources of amaranth in Mexico. Inifap, Central Regional Research Center, Celaya, Guanajuato, México.

13. Reyna Trujillo, Carmona Jiménez - Geographical Research, 1994 - Scielo.org.mx. Pluviometric characterization and distribution of amaranthus spp in México

14. FAO. 1990. Guide for pest management in underexploited andean crops. regional office for Latin America and the Caribbean, Santiago, Chile.

15. Kauffman, C. S., Hass N.N. Bailey, B.T. Volak, L.E. Weber and N.R. Volk. 1984. Amaranth grain production guide. rodale research report nc-83-6. Rodale press, inc., Emmaus, Pennsylvania, USA.

16. Kauffman, C.S., and L.E. Weber. 1990. grain amaranth. Janick and J.E. Simon (eds.), advances in new crops. timber press, Portland USA.

17. Weber, L.E., Kauffman, C.S.; Bailey, N.N. and b. t. Volak. 1985. Amaranth grain production guide. rodale press inc. Kutztown.

18. Cortes E. L. Espitia E., De la O M., Hernandez J. M., Bañuelos S. H., Ramírez m. l. 2010. Destination and uses of amaranth production in the high valleys of Mexico. in: xxxiii National and International Congress of Phytogenetics. Nuevo Vallarta, Nayarit.

19. Okuno, K. y S. Sakaguchi. 1981. Glutinous and non-glutinous starches in perisperm of grain amaranths. cereal res. commun. 9(4):305-310.

20.Lorenz, k. y f. Collins. 1981. Amarantus hypocondriacus -characteristics of the starch and baking potential of the flour-. starch/stärke 33:149-153.

21.Castilla F. 1982 Relationships of Amaranthus caudatus mp coons – Economic Botany, 1982 .

22.Bressani. R, Elijah. L.G., Garcia-Soto. A. limiting amino acid in amaranth grain protein from biological test. 1984

23.Acevedo et al., 1983. Badilla I. & Nobel, p: s. 1983. Water relations, diurnal acidity changes and productivity of a cultivated cactus, optunia, ficus-indica. palnt physiol.

24.Mendez-Llorente F.; Ramírez-Lozano, R. G.; Aguilera-Soto, j. i. and Arechiga-Flores, C. F. 2008. Performance and nutrient digestion of lambs fed incremental levels of wild cactus (opuntia leucotrichia). in: Conference on International research on food security, natural resource management and rural development University of Hohenheim.

25.Bravo Hollis, h. 1978. the cacti of México, vol. 1. UNAM, Mexico.

26.Garcia Corrales Joel., Nopalitos y tunas production, marketing, post-harvest and industrialization. Claudio a. Flores Valdez (editor) researchers of the nopal program of the center for economic, social and technological research of agroindustry and world agriculture (ciestaam), Universidad Autónoma Chapingo, first edition in spanish, 2003

27.El Kossori, r.l., Villanume, c., el Boustani, e., Sauvaire, & Mejean, l. 1998. Composition of pulp, skin and seeds of prickly pears fruit (opuntia ficus-indica sp.). Plant physiol., 72(3):775-780.

28.Trujillo 2009 "The cultivation of tuna" opuntia ficus indica. Regional Agrarian Management la Libertad, Peru.

29.Sánchez Mejorada, 1982, "Report on the Tucson meeting to analyze the cacti trade" in Mexico, mexican cacti and succulents.

30.National Commission for the knowledge and use of biodiversity. 2009 Mexico. nopales, prickly pears and xoconostles, mexican council of nopal and prickly pear, a.c./nopal rojo.